CW00570931

# SPIRAL SLIDE RULE.

**EQUIVALENT TO**

**A STRAIGHT SLIDE RULE 83 FEET 4 INCHES LONG,
OR, A CIRCULAR RULE 13 FEET 3 INCHES
IN DIAMETER.**

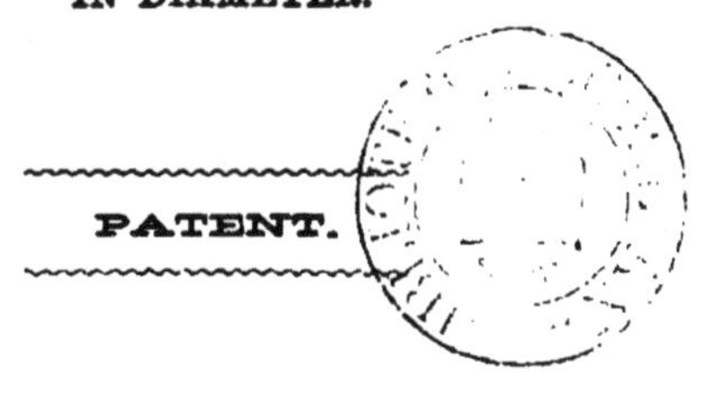

**PATENT.**


**GEORGE FULLER, M. INST. C.E.,**

PROFESSOR OF ENGINEERING IN THE QUEEN'S UNIVERSITY, IRELAND.

LONDON:

**E. & F. N. SPON, 46, CHARING CROSS.**

**NEW YORK: 446, BROOME STREET.**

**1878.**

# SPIRAL SLIDE RULE.

THE method of performing by mechanical means the work of addition and subtraction required when multiplying and dividing numbers by means of logarithms, originated with Gunter about the year 1606. He constructed a linear scale which was composed of two equal parts, and each half divided into parts proportional to the logarithms of numbers from 10 to 100. With this scale he used a pair of compasses for making the additions and subtractions.

About the year 1630, Oughtred invented the two similar logarithmic scales sliding in contact, which are at present in use; and he is stated to have used both straight and concentric circular scales. The advantage of this arrangement is, that the result is obtained by one motion of the sliding scale; and not only are multiplication and division thus worked, but questions in proportion, or the combination of the two, are solved by a single movement of the slide.

The simplicity of this method of calculating with figures is so great, that it seems strange it has not been more used; but the following considerations will, it is believed, account for this:—

In testing the relative advantages of different methods of making arithmetical calculations, the mental effort required, the time occupied, and the truth of the result have all to be taken into account. Now, in judging the ordinary slide rule by these points, it will be found that the facilities it offers are more apparent than real.

It is easy, with a little practice, to place one of the lines of the slide either opposite to a division of the rule or in a required position between two divisions, if these are not very close together. When the space, however, between two consecutive marks is very small, then great difficulty arises, from the strain upon the eyesight and the minute motion of the slide.

For example, in the ordinary slide rule with the scale $5\frac{1}{2}$ inches long, the breadth of the division from 99 to 100 is about $\frac{1}{40}$ of an inch. Therefore to mark such a number as 996, this space must be mentally divided into ten equal parts, each part consequently being $\frac{1}{400}$ of an inch, a magnitude quite inappreciable without a magnifying glass. The effort and time for the above is, however, slight, compared to that required when a point on one scale between two divisions has to be placed or read as agreeing with a point on the other, also between two divisions. For in this case (which is the most common, owing to the number of divisions on the ordinary slide rule necessarily being few) the division on one scale has to be mentally divided, and the particular point required fixed in the mind by its distance from the nearest division. Then the division on the other scale has to be mentally divided, and that part of it read which agrees with the point on the first scale previously fixed in the mind. Thus, for example, suppose it is required to place 554 on one scale to agree with 643 on the other. There are marks at 55 and 56 on one scale, and at 64 and 65 on the other; but the $\frac{4}{10}$ part of the distance between 55 and 56 has to be made to coincide with the $\frac{3}{10}$ part of that between 64 and 65: the difficulty not being to divide either of these distances into ten parts, if they are not very small, but to combine the two operations together.

If at the same time the spaces between the marks are very small, the difficulty is greatly increased by the strain upon the eyesight.

With regard to the truth of the result, Mr. Heather, in his 'Treatise on Mathematical Instruments,' writes in relation to the foot slide rule: "The solution in fact may be considered as obtained to within a two-hundredth part of the whole." Now this approximation, though close considering the length of scale of the instrument, and sufficient for some, is not near enough for very many of the calculations required by engineers and architects.

From the above it appears that a slide rule to

be thoroughly efficient, so that calculations may be made by it with ease and rapidity, and practically correct results obtained, the length of the logarithmic scale should be such that the space between any two consecutive numbers is large enough to be easily distinguished by the unaided eye; that the scale should be read by indices, and not as in the present rules; and that the number of divisions should be so great and distinctly marked, that the result to be obtained may be easily read and practically correct.

This combination, it is believed, is attained in the spiral slide rule.

The rule consists of a cylinder (*d*) that can be moved up and down upon, and turned round, an axis (*f*), which is held by a handle (*e*). Upon this cylinder is wound in a spiral a single logarithmic scale. Fixed to the handle is an index (*b*). Two other indices (*c*) and (*a*), whose distance apart is the axial length of the complete spiral, are fixed to the cylinder (*g*). This cylinder slides in (*f*) like a telescope tube, and thus enables the operator to place these indices in any required position relative to (*d*). Two stops (*o*) and (*p*) are so fixed that when they are brought in contact, the index (*b*) points to the commencement of the scale. (*n*) and (*m*) are two scales, the one on the piece carrying the movable indices, the other on the cylinder (*d*).

It will at once be seen that by this arrangement the length of the logarithmic scale can be

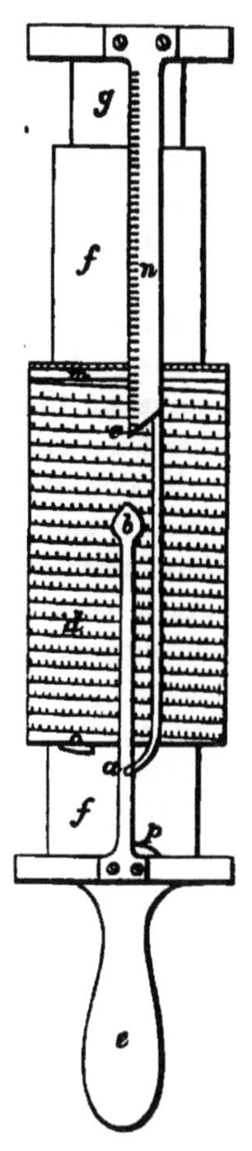


Scale, 3 inches to a foot.

made very great, whilst keeping the instrument of a convenient size for use. It requires only one logarithmic scale, so that every inch of the spiral scale is equivalent to two of the ordinary straight rule.

To fulfil with great exactness all the necessary conditions, the scale is made 500 inches, or 41 feet 8 inches, long—equivalent therefore to a straight rule 83 feet 4 inches long, and to a circular rule 13 feet 3 inches in diameter. This allows of results being obtained to one ten thousandth part of the whole, at the same time requiring no space less than $\frac{1}{48}$ of an inch, between any two consecutive numbers of four figures. The length of scale in the common rule only permits of the *first* figure of a number being printed; in this rule, the first *three* figures are printed throughout the scale. With this rule, to produce an error of one part in 200, there must be, either in setting or reading, an error of *one and one-tenth* inches. It will also be seen that both for setting and reading, indices are applied to the scale; so that these operations are performed with the greatest mental ease.

It may be remarked that the slide rule possesses an advantage over a table of logarithms, in addition to that of performing mechanically the requisite additions and subtractions, in that the approximation is uniform throughout the scale, and not nearer in one part than in another, as in the tables.

It must be remembered that all the calculations

founded on measurements of length, weight, and time, can only be approximative, as the data for them are so. Except therefore with the most refined measurements, it is a waste of time to carry results beyond the ten thousandth part of the whole.

## RULES.

In using the slide rule, the handle should be held in the left hand, the movable cylinder and indices being worked by the right, which holds the pen or pencil.

*Division of the Scale.*—Though this scale is large enough to admit of being read to four or even five figures, space does not allow of its being figured to more than three. Each of the primary divisions, as far as 650, is divided into ten parts, and from thence to 1000 into five parts; so that all numbers of four figures have either a mark upon the scale, or are midway between two marks. Thus 4786 is shown by a mark; also 8432; but 8431 is not shown by a mark, but is midway between 8430 and 8432. In a large part of the scale the space between these secondary divisions is large enough to be easily divided into parts by the eye. Thus many numbers of five figures are easily shown; for example, 26854. There are the first three figures at 268, then 5 is at the fifth secondary division, and the 4 must be estimated by the eye as $\frac{4}{10}$ of the space between 2685 and 2686. It must be noted

that the same figures do not always mea
same amount. Thus to represent 268540, :
2685·4, 268·54, 26·854, 2·6854, ·26854, ·0:
·0026854, &c., the same point on the scale is

MULTIPLICATION.

*Rule.*—Bring 100 to the fixed index, and
the movable index to the multiplicand.
move the cylinder so that the multiplier is
fixed index. The quotient is read off at
the movable indices.

*To ascertain the Value of the Quotient.*—Mo
quently this may be determined by inspectio
following rules will, however, give it in all ca

Consider a number like—

| | | | |
|---|---|---|---|
| 18763 | as one of | 5 | figures |
| 1876·3 | ,, | 4 | ,, |
| 187·63 | ,, | 3 | ,, |
| 18·763 | ,, | 2 | ,, |
| 1·8763 | ,, | 1 | ,, |
| ·18763 | ,, | 0 | ,, |
| ·018763 | ,, | −1 | ,, |
| ·0018763 | ,, | −2 | ,, |
| ·000187 | ,, | −3 | ,, |

Then the number of figures in the quot
the algebraic sum of the number of figures
multiplier and multiplicand, if it is *not* reac
the same index as the multiplicand. It is c
than that sum if read upon the same index.

*Examples.*—12 × ·142 = 1·704. The s

figures is two, and the answer is read from the same index as the multiplicand, so that the quotient has one figure.

64 × ·24 = 15·36. The sum of figures is two, but the answer is read at a different index to the multiplicand, and therefore the quotient has two figures.

12 × ·00142 = ·01704. The sum of the figure is 0, but the answer is read at the same index with the multiplicand, and therefore the quotient has one less, or minus one.

64 × ·0024 = ·1536. The sum of the figures is 0, and the answer is read at a different index to the multiplicand, and therefore the quotient has 0 figures.

48·42 × 64·34 = 3115·3. The sum of the figures is 4, and the answer is read at a different index to the multiplicand, and therefore the quotient has four figures.

DIVISION.

*Rule.*—Place *divisor* to fixed index, and the upper or lower movable index to the dividend, according as the first figure in the divisor is greater or less than the first figure in the dividend. Then move the cylinder so that the fixed index is at 100, and read the quotient at one of the movable indices.

The number of figures in the quotient is the algebraic *difference* between the number of figures in the dividend and divisor, if it is *not* read upon the same index as the dividend. It is one *more* than that difference if read upon the same index.

*Examples.*—1468 ÷ 63 = 23·3, as the difference is 2, and the quotient is *not* read upon the same index as the dividend.

1468 ÷ 125 = 11·7, as the difference is 1, and the quotient *is* read upon the same index as the dividend, and therefore has *two* figures.

·1468 ÷ 63 = ·00233, as the difference is − 2, and the quotient is *not* read upon the same index as the dividend, and therefore has − 2 figures.

1468 ÷ ·00125 = 1174000, as the difference is 4 − (− 2) = 4 + 2 = 6, and the quotient *is* read upon the same index as the dividend, and therefore has *seven* figures.

## Multiplication and Division.

*Rule.*—Move the cylinder so as to place the denominator to the fixed index. Then place movable index to one of the numerators. Then move the cylinder so that the fixed index points to the other numerator, and read the quotient at one of movable indices.

The number of figures in the quotient is the algebraic difference between the sum of the number of figures in the numerator and in the denominator, if it is read upon the *same* index as a factor of the numerator. It is *one more* than that difference if read upon the other index.

*Example.* $\frac{4854 \times 32 \cdot 6}{536} = 295 \cdot 22$, as the difference is $4 + 2 - 3 = 3$, and the quotient is read at the same index as either 4854 or 32·6 is placed.

$\frac{\cdot 0764 \times \cdot 032}{14 \cdot 63} = \cdot 000167$, as the difference is $(-1) + (-1) - 2 = -4$, and the quotient is not read upon the index, that either ·0764 or ·032 is placed, and therefore *the number* of figures is $-3$.

To multiply three numbers together when one of them is a constant in frequent use.

*Rule.*—Find the *reciprocal* of the constant by division, and use it as the divisor in the preceding rule.

### Ratio.

When either of the movable indices is at one number and the fixed index at another, and the cylinder is turned into any other position, though the numbers at the indices will be different, their ratio will remain constant.

*Example.*—To convert francs and centimes into sterling money, supposing exchange 25f. 45c. for 1*l.* The ratio between centimes and pence is 2545 to 240. Place the cylinder so that the fixed index is at 2545, and make one of the movable indices point to 240. Then on moving the cylinder to read off different numbers of centimes at the fixed index, the corresponding value in pence will be read at the movable index.

*Wages Table.*—To find the wages for different times at 35*s.* per week of 57 hours. Place the

cylinder so that the fixed index is at 57, and make one of the movable indices point to 420, the number of pence in 35*s*. Then on moving the cylinder to read off different numbers of hours at the fixed index, the corresponding wages in pence will be read at the movable index.

## Proportion.

To find a third proportion to two numbers— $a : b :: b : c$

$c = \frac{b \times b}{a}$. Proceed according to rule for multiplication and division.

To find a fourth proportion to three numbers— $a : b :: c : d$

$d = \frac{b \times c}{a}$. Proceed according to rule for multiplication and division.

## Powers and Roots.

To obtain the square, cube, and fourth power of a number. The quickest way with this rule is by direct multiplication.

For higher powers and roots. Place the upper movable index ($c$) to the number, and read the scales ($n$ and $m$). These together give the *mantissa* of the logarithm of the number. To this the *index* has to be added. The index of the logarithm of a number greater than unity is *one less* than the number of figures in the integral part of that number. Thus the index of 5432 is 3, of 543·2 is 2, of 54·32 is 1, and of 5·432 is 0.

Multiply or divide the resulting number by the power or root, as shown above. Then place the cylinder so that it reads on the scales (*n* and *m*) the decimal part of the quotient. The power or root is then at the index (*c*). In the result the number of figures before the decimal point is *one more* than the number in the integral part of the above quotient.

The scale (*n*) is read from the *lowest line* of the top spiral and (*m*) from the vertical edge of the scale (*n*).

*Examples.*—$5^{13}$, on placing (*c*) to 500, scale (*n*) reads ·68 and scale (*m*) ·01897, which gives the logarithm of 5 – ·69897, the index being 0. Then ·69897 × 13 = 9·08661. Now placing the cylinder so that it reads ·08661 on scales (*n* and *m*) the index (*c*) reads 12207, and the required power is 1220700000, having 10 figures, as the integral part of the above quotient is 9.

$\sqrt[5]{741}$ on placing (*c*) to 741, scale (*n*) reads ·86 and scale (*m*) ·00982 which gives the logarithm of 741 – 2·86982, the index being 2. Then 2·86982 ÷ 5 = ·57396. Now placing the cylinder so that it reads ·57396 on scales (*n* and *m*) the index (*c*) reads 37495, and the required root is 3·7495, having one figure before the decimal point, as the integral part of the above quotient is 0.

### Powers of Decimal Fractions.

To avoid the use of negative indices, which often lead to erroneous results unless they are frequently used, the following method may be adopted :—

Write them as vulgar fractions, the numerator being expressed in units and decimals, and raise the numerator and denominator to the required power, the former by the method given above; the latter can be written down at once.

$$\text{Thus } \cdot 47^3 = \left(\frac{4 \cdot 7}{10}\right)^3 \qquad \cdot 047^3 = \left(\frac{4 \cdot 7}{100}\right)^3$$

### Roots of Decimal Fractions.

Write them as vulgar fractions, and multiply numerator and denominator by ten or a power of ten, so that the denominator may have a complete root. Then take the required root of the numerator by the method given above, and of the denominator by inspection :

$$\text{Thus } \sqrt{\cdot 4} = \sqrt{\frac{4}{10}} = \sqrt{\frac{40}{10^2}} = \frac{\sqrt{40}}{10}$$

$$\sqrt[3]{\cdot 04} = \sqrt[3]{\frac{4}{10^2}} = \sqrt[3]{\frac{40}{10^3}} = \frac{\sqrt[3]{40}}{10}$$

$$\sqrt[5]{\cdot 586} = \sqrt[5]{\frac{586}{10^3}} = \sqrt[5]{\frac{58600}{10^5}} = \frac{\sqrt[5]{58600}}{10}$$

$$\sqrt[3]{\cdot 00065} = \sqrt[3]{\frac{65}{10^5}} = \sqrt[3]{\frac{650}{10^6}} = \frac{\sqrt[3]{650}}{10^2}$$

$$(\cdot 0434)^{\frac{5}{6}} = \left(\frac{434}{10^4}\right)^{\frac{5}{6}} = \left(\frac{43400}{10^6}\right)^{\frac{5}{6}} = \frac{(43400)^{\frac{5}{6}}}{10^5}$$

## Simple Interest.

Let P be the Principal in pounds and parts of a pound;

$n$ the number of years and parts of a year for which interest is taken;

$r$ the interest of one pound for one year;

M the amount.

$M = P + P\,n\,r.$

Also P is the present value of M, due at the end of the time $n$.

$$P = \frac{M}{1 + n\,r}.$$

In practice the ***discount*** is the interest of the sum of money paid before it is due;

$$\text{or } D = M\,n\,r.$$

## Compound Interest.

Let P be the Principal in pounds and parts of a pound;

$n$ number of years for which interest is taken;

$r$ the interest of one pound for one year;

M the amount.

Interest due once a year.

$$M = P\,(1 + r)^n \qquad n = \frac{\log. M - \log. P}{\log. (1 + r)}$$

Let interest be due $q$ times a year and $\frac{r}{q}$ the interest of one pound for $\frac{1}{q}$ part of a year,

$$M = P\left(1 + \frac{r}{q}\right)^{qn}.$$

VALUE OF *r*.

| Per Cent. | | Per Cent. | | Per Cent. | | Per Cent. | | Per Cent. | |
|---|---|---|---|---|---|---|---|---|---|
| $\frac{1}{16}$ | ·000625 | $\frac{3}{8}$ | ·00375 | $\frac{11}{16}$ | ·006875 | 1 | ·01 | 6 | ·0( |
| $\frac{1}{8}$ | ·00125 | $\frac{7}{16}$ | ·004375 | $\frac{3}{4}$ | ·0075 | 2 | ·02 | 7 | ·0 |
| $\frac{3}{16}$ | ·001875 | $\frac{1}{2}$ | ·005 | $\frac{13}{16}$ | ·008125 | 3 | ·03 | 8 | ·0 |
| $\frac{1}{4}$ | ·0025 | $\frac{9}{16}$ | ·005625 | $\frac{7}{8}$ | ·00875 | 4 | ·04 | 9 | ·( |
| $\frac{5}{16}$ | ·003125 | $\frac{5}{8}$ | ·00625 | $\frac{15}{16}$ | ·009375 | 5 | ·05 | 10 | · |

The tables printed on the rule have been ma
and selected as those considered most usef
Owing to our want of a decimal system, it l
been deemed most important to have a serie
of tables which give for our measures of weigh
length, time, &c., the equivalent decimal fract
of the larger for successive numbers of tl
smaller unit. This enables results to be obtain
without the necessity of reduction. Thus to fir
the area of a rectangle whose sides are 24′ 6
and 43′ 5½″. The table gives by inspect
·5208 and ·4583 opposite 6¼″ and 5½″ resp
tively, so that the area is obtained by multiplyi
24·521 by 43·458. The result, as shown by t
rule, is 1065·6. If the parts of a square f
are required in twelfths, the table shows th
·6 of a foot is equivalent to 7¼ twelfths, and tl
result reads 1065 – 7¼.

LONDON: PRINTED BY WILLIAM CLOWES AND SONS, STAMFORD STREE
AND CHARING CROSS.

Milton Keynes UK
Ingram Content Group UK Ltd.
UKHW022031190124
436367UK00004B/235